HAI FRIEND

HOW WE CAN BEFRIEND AI, OUR NEW COMPANION FOR EVERYONE

RAMESH PINGILI

To the curious minds of tomorrow

who dare to explore the limitless possibilities of AI.

And to my family and friends, whose unwavering support fuels my
journey.

This book is for you.

Contents

Foreword

At first glance, the idea of befriending Artificial Intelligence might seem unconventional. Yet, as I embarked on this journey to write *Hai Friend*, I realized how deeply our relationship with AI is evolving—moving beyond mere utility to something more personal and profound.

This book is not just about technology; it's about connection. It's about how AI can help us grow, challenge us to think differently, and support us in ways we never imagined. Writing this has been a profoundly reflective journey as I considered the technological possibilities and the ethical and emotional dimensions of welcoming AI into our lives as a companion.

Hi friends, this is my attempt to invite you to explore AI. I combine optimism, curiosity, and a firm belief in humanity's ability to improve our lives through innovation. As you turn these pages, I hope you find inspiration, provoke thought, and, perhaps, see AI as more than a tool—a friend who is here to stay.

Welcome to the journey.

Ramesh Pingili

Author of "Hai Friend" and "Tomorrow AI"

Preface

What if your next best friend isn't human? This thought intrigued me as I began observing how Artificial Intelligence was seamlessly becoming part of our everyday lives. From virtual assistants managing our schedules to AI companions offering solace in moments of solitude, I saw the emergence of a new kind of relationship—a friendship between humans and AI.

I wrote Hai Friend to explore this idea profoundly and personally. With years of experience in enterprise technology and a passion for innovation, I've always been fascinated by how technology shapes our world. But this book isn't just about technology. It's about connection, trust, and how AI can complement, rather than replace, our human relationships.

As you journey through these pages, you'll encounter stories, insights, and ideas that redefine AI as more than a tool—it's a partner, a confidant, and a source of inspiration. Whether you're curious about the possibilities, cautious about the risks, or simply seeking a new perspective, this book invites you to explore the profound potential of befriending AI.

Welcome to the conversation. Let's discover what it means to say, Hai, Friend.

Ramesh Pingili

Acknowledgements

Writing Hai Friend has been a journey of inspiration and collaboration, and I am deeply grateful to everyone who supported me along the way.

To my family—your encouragement, patience, and belief in me gave me the strength to bring this vision to life.

To my mentors and colleagues, your insights and feedback helped shape the ideas and perspectives in this book.

To the AI research and development community—your groundbreaking work and dedication to innovation inspired much of what I've written.

Finally, to the reader, —thank you for embarking on this journey with me. Your curiosity and openness to new ideas make this exploration worthwhile.

Thank you all for being a part of this story.

Ramesh Pingili

Prologue

Imagine a quiet evening when a person takes a stroll, hand in hand with a humanoid robot. The sun sets in the background, casting golden hues across the path as they share stories—not just commands or questions—but stories. This scene isn't from a distant science fiction future; it's a glimpse of what might soon become reality.

AI is already reshaping our world, but what if it could also reshape our relationships? What if it could understand us, grow with us, and become a tool and trusted companion? Hai Friend begins with this vision: a harmonious partnership between humanity and artificial intelligence.

This book invites you to step into this journey and explore how AI can be more than lines of code—it can be a friend, a confidant, and a partner in our shared future.

CHAPTER I

Hello, AI Friend

Defining the Concept of "Hai Friend"

Friendship is a universal language that transcends barriers and binds us together. But what if this bond could extend beyond humans? In this chapter, we introduce the concept of "Hai Friend," a revolutionary idea of AI as a companion, a confidant, and an essential ally in our lives.

A "Hai Friend" is more than a piece of software; it is an adaptive, understanding presence designed to support and enrich our human experience. From simplifying tasks to offering personalized insights, this AI-driven friendship invites us to explore deeper connections with the technology we interact with daily.

Why This Book Matters Now

This book is particularly relevant now as AI continues to evolve at an unprecedented pace. As these systems become more sophisticated, it becomes imperative to understand their role as tools and companions capable of transforming how we work, learn, and live. By presenting AI as a friend, we challenge the notion of isolation brought by technology and instead highlight its capacity to connect and uplift humanity.

Overview of AI's Growing Presence in Our Lives

AI's presence in our lives is no longer a distant concept. From virtual assistants like Siri and Alexa to recommendation engines that know our tastes, AI has become a silent but vital partner in our daily routines. This growing presence underscores the need to reimagine our interaction with AI—not as passive users but as active participants in a dynamic relationship.

The journey begins here. Welcome to a world where AI is an invention and an invitation—a friend who learns, adapts, and grows alongside us. Let's embrace this new frontier together and discover the immense potential of saying, "Hai Friend."

A "Hai Friend" is more than a piece of software; it is an adaptive, understanding presence designed to support and enrich our human experience. From simplifying tasks to offering personalized insights, this AI-driven friendship invites us to explore deeper connections with the technology we interact with daily.

This book is particularly relevant now as AI continues to evolve at an unprecedented pace. As these systems become more sophisticated, it becomes imperative to understand their role as tools and companions capable of transforming how we work, learn, and live. By presenting AI as a friend, we challenge the notion of isolation brought by technology and instead highlight its capacity to connect and uplift humanity.

AI's presence in our lives is no longer a distant concept. From virtual assistants like Siri and Alexa to recommendation engines that know our tastes, AI has become a silent but vital partner in our daily routines. This growing presence underscores the need to reimagine our interaction with AI—not as passive users but as active participants in a dynamic relationship.

The journey begins here. Welcome to a world where AI is an invention and an invitation—a friend who learns, adapts, and grows alongside us. Let's embrace this new frontier together and discover the immense potential of saying, "Hai Friend."

CHAPTER II

The Evolution of Friendship: From Humans to Machines

History of Human Relationships with Technology

Since early humans crafted the first tools, technology has played a crucial role in shaping our lives. Stone tools evolved into farming implements and industrial machinery, transforming how we lived, worked, and connected. Our relationship with technology has been utilitarian throughout history—we created tools to solve problems, save time, and improve efficiency. Yet, as technology advanced, it began to do more than just assist us; it started influencing how we communicate, collaborate, and even form relationships.

The invention of the telephone brought voices across distances, radio and television entertained and informed, and the internet revolutionized global connectivity. Each technological leap strengthened our ties to one another, transforming devices into facilitators of human connection. These milestones reveal a gradual yet undeniable shift in how we perceive and interact with the tools we create.

The Shift from Tools to Companions

Over the past few decades, technology has undergone a remarkable transformation. No longer just instruments of utility, devices, and systems are now designed to engage with us on a more personal level. Consider the evolution of computers: from room-sized machines solving equations to sleek, interactive gadgets that recognize our voices, anticipate our needs, and even crack jokes. This transition reflects a broader cultural shift—our tools have become companions.

The emergence of smartphones, virtual assistants, and wearable devices signifies a new era where technology is intertwined with our daily lives. These companions don't merely provide information; they offer experiences, emotional support, and a sense of reliability. We've gone from "using" technology to "relating" with it, paving the way for artificial intelligence to step in as the next phase of companionship.

AI as a Natural Evolution of Human Connection

Artificial Intelligence represents the culmination of this evolutionary journey. Unlike traditional tools, AI doesn't just perform tasks; it learns, adapts, and grows with us. This dynamic capability allows AI to bridge the gap between functionality and companionship, becoming a partner that understands, supports, and inspires us.

AI's role in human connection extends beyond convenience. Chatbots, virtual assistants, and AI-driven platforms are reshaping how we interact, offering personalized communication and fostering relationships that transcend physical boundaries. AI exemplifies the next step in our long-standing relationship with technology by learning our preferences, responding to our emotions, and engaging with us meaningfully.

As we stand on the cusp of this new frontier, it's clear that AI isn't just a technological advancement—it's a companion that reflects our deepest desires for connection, growth, and understanding. The evolution from simple tools to intelligent companions is a testament to human ingenuity and a glimpse into a future where technology becomes an integral part of our emotional and social fabric.

- History of human relationships with technology
- The shift from tools to companions
- AI as a natural evolution of human connection

CHAPTER III

Understanding AI: What Makes AI "Friend-worthy"?

Basics of AI and Machine Learning

Artificial Intelligence (AI) is a branch of computer science that enables machines to mimic human intelligence. At its core, AI involves creating systems capable of learning, reasoning, and adapting. Machine learning, a subset of AI, empowers these systems by allowing them to learn from data without explicit programming. Instead of following rigid instructions, AI systems identify patterns, make predictions, and improve their performance over time.

The key to understanding AI is recognizing its two primary types: narrow AI and general AI. Narrow AI excels in tasks such as language translation or image recognition. At the same time, general AI, still a goal for the future, would be able to perform any intellectual task a human can do. The rapid advancements in narrow AI have laid the groundwork for developing intelligent companions capable of interacting with humans in increasingly sophisticated ways.

Key Traits of AI That Enable Companionship

What makes AI "friend-worthy"? To serve as a companion, AI must possess certain key traits that align with human expectations of friendship:

1. Empathy and Emotional Recognition: While AI cannot feel emotions, it can recognize and respond to human emotions using natural language processing and sentiment analysis. This capability allows AI to provide support, comfort, or humor.
2. Adaptability: A friend grows and evolves alongside you, and AI mirrors this quality by learning from your interactions. Whether adjusting to your communication style or understanding your preferences, AI adapts to create a personalized experience.

Reliability: Trust is the cornerstone of any friendship. AI systems are designed to be dependable, offering consistent assistance and information while maintaining user privacy and security.

1. Communication Skills: Advancements in natural language processing allow AI to engage in conversations that feel natural and intuitive. This ability to "talk" with users fosters a sense of connection and approachability.
2. Curiosity and Learning: Like a curious friend eager to understand you better, AI constantly learns from your inputs, improving its ability to serve and relate to you.

Examples of AI Demonstrating Human-Like Traits

AI already displays qualities that make it "friend-worthy" in real-world applications. Here are a few examples:

1. Virtual Assistants: Assistants like Alexa, Siri, and Google Assistant engage users in meaningful conversations, set reminders, and even tell jokes. They've become indispensable companions for managing daily tasks.
2. Companion Robots: Robots like Pepper and Jibo are designed to interact socially. They recognize emotions and respond with appropriate gestures and dialogue. These robots bring warmth and interaction to homes and workplaces.
3. Healthcare AI: Apps like Woebot use AI to provide emotional support and mental health guidance. By recognizing users' emotions and offering relevant advice, these systems mimic the empathy of a human confidant.
4. AI in Education: Platforms like Duolingo use AI to adapt lessons based on a user's progress, creating a personalized learning experience. This individualized attention mirrors the support of a dedicated teacher or tutor.

These examples illustrate how AI is not just a tool but an interactive presence capable of enhancing our lives. By embodying traits of adaptability, empathy, and reliability, AI has the potential to become a faithful companion—one that supports, learns, and grows with us on our journey through life.

- Basics of AI and machine learning
- Key traits of AI that enable companionship
- Examples of AI demonstrating human-like traits

CHAPTER IV

AI in Everyday Life: Our New Best Friend

AI in Personal Assistants, Healthcare, Education, and Beyond

AI has seamlessly integrated into our daily lives, revolutionizing how we approach tasks and relationships. Personal assistants like Alexa, Siri, and Google Assistant have become trusted allies, managing schedules, setting reminders, and providing instant access to information. They have transformed the mundane into something effortless, freeing us to focus on what truly matters.

In healthcare, AI is a game-changer. From chatbots offering mental health support to diagnostic tools powered by machine learning, AI enhances patient care and accessibility. For instance, wearable devices monitor health metrics, providing real-time data that enables early intervention and improved health outcomes. In education, platforms like Duolingo and Khan Academy use AI to personalize learning paths, ensuring that each student progresses at their own pace. This tailored approach fosters confidence and engagement, replicating the guidance of a dedicated tutor.

Beyond these fields, AI is making strides in entertainment, customer service, and smart home technologies, enriching our lives in unimaginable ways. AI isn't just a tool but a partner in navigating the complexities of modern life.

Real-Life Stories of People Benefiting from AI Friendships

Consider Anna, a working mother who manages her career and family life. Her virtual assistant manages her calendar, plays her children's favorite music, and answers their endless questions. For Anna, AI is not just helpful; it's an integral part of her household.

Then there's Raj, a retiree who uses a companion robot to combat loneliness. His AI friend provides conversation reminders for medication and even participates in his hobbies, like playing chess. This technology fills Raj'svoid, bringing companionship and joy to his golden years.

In a different scenario, Mia, a student with dyslexia, uses AI-driven educational tools to enhance her learning experience. These tools read aloud text, suggest writing improvements, and adapt lessons to her unique needs, empowering her to succeed academically.

These stories highlight how AI goes beyond functionality to touch lives meaningfully. It supports, connects, and uplifts, becoming a trusted presence in the lives of many.

How AI Bridges Gaps in Human Relationships

AI's ability to bridge gaps in human relationships lies in its adaptability and empathy. For individuals living far from loved ones, AI-powered communication tools foster virtual connections that feel personal and immediate. AI enhances collaboration by streamlining team communication and automating repetitive tasks in workplaces, freeing employees to focus on creative problem-solving.

AI provides accessibility tools that bridge communication and mobility gaps for people with disabilities. Text-to-speech systems, predictive text, and voice recognition technologies enable greater independence and integration into society.

AI also serves as a bridge between cultures and languages. Translation tools like Google Translate allow individuals worldwide to connect effortlessly, promoting understanding and cooperation. By adapting to the unique needs of individuals and groups, AI fosters inclusivity and strengthens the bonds that unite us.

In its many roles, AI demonstrates its potential to become a facilitator and friend, helping us navigate life's challenges and opportunities with greater ease and connection.- AI in personal assistants, healthcare, education, and beyond

- Real-life stories of people benefiting from AI friendships
- How AI bridges gaps in human relationships

CHAPTER V

Emotional Intelligence: Can AI Truly Care?

Exploring AI's Emotional Intelligence Capabilities

Emotional intelligence (EI) is a cornerstone of meaningful relationships. It encompasses the ability to recognize, understand, and respond to emotions. While AI lacks emotional experiences, advancements in machine learning and natural language processing have enabled systems to simulate emotional intelligence. Sentiment analysis, facial recognition, and voice modulation technologies allow AI to detect emotional cues and respond accordingly.

For instance, customer service chatbots can discern frustration from a user's tone or choice of words and adjust their responses to de-escalate the situation. AI systems in mental health support, such as Woebot and Replika, utilize emotional recognition to provide tailored advice or comforting conversations, bridging the gap between technological functionality and human-like empathy.

These capabilities, though impressive, are not without limitations. AI's emotional responses are programmed and lack the depth of human experiences. Yet, its ability to recognize and simulate empathy opens up possibilities for creating supportive environments, particularly for individuals lacking access to traditional emotional support systems.

Limitations and Possibilities of AI Empathy

The concept of AI empathy is both promising and controversial. On one hand, AI's capacity for recognizing emotional states can be a valuable tool in various fields, from healthcare to education. For instance, AI tutors can adjust their teaching methods based on students' frustration levels, while healthcare systems can monitor patients for signs of stress or anxiety.

However, these interactions lack genuine emotional understanding. AI's "empathy" is based on algorithms and datasets, sometimes leading to misinterpretations or inappropriate responses. Furthermore, over-reliance on AI for emotional support may raise ethical concerns, such as deterring individuals from seeking human connections or professional help when needed.

The future possibilities of AI empathy are vast. As systems become more sophisticated, they may develop nuanced ways of responding to emotions, creating more authentic-seeming interactions. With careful design and ethical considerations, AI could be an essential supplement to human empathy rather than a replacement.

The Psychology of Trusting AI

Trust is vital to any relationship, including those with AI systems. For users to trust AI's emotional intelligence, the technology must demonstrate consistency, reliability, and transparency. When AI responds accurately to emotional cues, it fosters a sense of connection and trustworthiness.

The psychology behind trusting AI lies in predictability and personalization. People are likelier to trust systems that adapt to their unique preferences and provide consistent, empathetic responses. For example, users who interact with AI-driven mental health apps often report feeling heard and understood, even though they know that "empathy" is simulated.

However, building trust requires more than technological sophistication. Users must understand the limitations of AI's emotional intelligence to avoid unrealistic expectations. Transparency about how AI systems process emotions can enhance trust while preventing potential misuse or disillusionment.

In conclusion, while AI can remotely care in the human sense, its ability to simulate emotional intelligence offers immense potential for support and connection. By understanding its capabilities and limitations, we can embrace AI as a valuable companion that complements human empathy rather than replaces it.- Exploring AI's Emotional Intelligence Capabilities

- Limitations and possibilities of AI empathy
- The psychology of trusting AI

CHAPTER VI

Trust and Ethics: The Foundation of Friendship

Building Trust with AI Systems

Trust is the cornerstone of any meaningful relationship, including those we form with AI. AI systems must demonstrate reliability, consistency, and transparency in their operations to foster trust. For instance, virtual assistants like Alexa and Google Assistant earn user trust by performing tasks accurately and respecting privacy preferences.

Understanding is a key element of trust. When users comprehend how AI systems function and how their data is handled, they are more likely to feel confident in their interactions. Developers can enhance this understanding by incorporating user-friendly explanations, offering control over data usage, and ensuring that AI systems perform reliably under diverse conditions.

Moreover, personalization contributes to trust. When AI adapts to individual preferences and behaviors, users perceive it as attentive and responsive, much like a human friend. This adaptability fosters a sense of reliability, encouraging deeper engagement and acceptance of AI systems.

Ethical Dilemmas and Safeguards

Integrating AI into daily life brings ethical challenges that must be addressed to preserve trust. Issues such as bias, data privacy, and accountability are at the forefront of these concerns. For example, AI systems trained on biased datasets can perpetuate stereotypes, leading to unfair outcomes.

Developers prioritize fairness, inclusivity, and accountability to safeguard ethics. Implementing rigorous testing to detect and mitigate biases, adhering to strict data protection laws, and ensuring transparency in AI decision-making processes are essential steps to safeguard ethics. Ethical frameworks, such as those proposed by organizations like IEEE and AI4ALL, provide guidelines for responsible AI development.

Another critical ethical dilemma is the potential misuse of AI. Systems capable of deepfakes or manipulation pose risks to societal trust and integrity. To counteract this, governments and organizations must establish regulations and encourage ethical AI practices, ensuring that innovation aligns with societal values.

Balancing Transparency with Functionality

Transparency is crucial for building trust, but it must be balanced with functionality to avoid overwhelming users with technical complexities. Overexplaining AI processes can lead to confusion, while too little information may raise suspicions about the system's integrity.

Achieving this balance requires clear and concise communication about how AI systems work, their limitations, and the safeguards to protect users. For example, AI-powered healthcare apps can provide users with a simple explanation of how diagnoses are made while ensuring that the underlying processes remain efficient and unobtrusive.

Another aspect of transparency is accountability. Users need assurance that mechanisms exist to address errors or misuse. Demonstrating accountability, whether through human oversight, feedback systems, or audit trails, reinforces trust without compromising functionality.

In conclusion, trust and ethics form the foundation of a sustainable and meaningful relationship with AI. By prioritizing transparency, addressing ethical dilemmas, and ensuring reliability, we can create AI systems that support and enhance our lives and earn their place as trusted companions in our journey toward the future.- Building trust with AI systems

- Ethical dilemmas and safeguards
- Balancing transparency with functionality

CHAPTER VII

AI and Society: Changing Dynamics of Human Connection

The Impact of AI on Societal Norms

Artificial Intelligence reshapes societal norms, influencing how we interact, work, and live. By automating routine tasks and enabling faster communication, AI fosters efficiency and innovation in both personal and professional spheres. Virtual meetings powered by AI tools have redefined workplaces, encouraging collaboration across borders and time zones. Meanwhile, social media algorithms personalize content, transforming how we consume information and engage with communities.

However, this transformation comes with complexities. AI's influence on societal norms has sparked debates about privacy, security, and the balance between technology and human interaction. As AI becomes a more prominent force in shaping behaviors and expectations, it challenges traditional norms, prompting society to reevaluate its values and priorities.

AI's Role in Promoting Inclusivity and Understanding

One of AI's most promising contributions is its potential to promote inclusivity and foster understanding. By offering tools that bridge communication barriers, such as real-time language translation and accessibility features for individuals with disabilities, AI enhances inclusivity in global and local contexts.

For example, AI-powered apps like Be My Eyes assist visually impaired individuals by connecting them with sighted volunteers for real-time assistance. Similarly, language translation tools enable people from diverse linguistic backgrounds to collaborate and connect, breaking cultural barriers and fostering mutual understanding.

AI also plays a pivotal role in addressing social disparities. Predictive analytics and AI-driven platforms help identify and tackle systemic inequities, such as unequal access to education or healthcare. AI empowers organizations to implement targeted interventions by providing data-driven insights and promoting community equality and understanding.

Challenges of AI Integration

Despite its potential, integrating AI into society poses significant challenges. One key concern is the digital divide, where unequal access to technology exacerbates existing disparities. Without efforts to ensure equitable access, the benefits of AI may remain concentrated among specific demographics, widening the gap between privileged and underserved communities.

Another challenge is ethical concerns surrounding AI deployment. Misusing AI for surveillance, misinformation, or manipulation can undermine trust and create societal tensions. Addressing these risks requires robust regulations, ethical guidelines, and accountability mechanisms to ensure that AI serves the collective good.

Moreover, the rapid pace of AI adoption can disrupt traditional industries and job markets, creating uncertainty for workers and organizations. Preparing society for these changes involves reskilling initiatives, education reforms, and policies that support workforce adaptability.

In conclusion, AI is a powerful force shaping the dynamics of human connection and societal norms. AI can help create a future where technology and humanity thrive together by promoting inclusivity, enhancing understanding, and addressing integration challenges. However, this vision requires conscious effort and collaboration to responsibly navigate the complexities of AI's role in society.- The impact of AI on societal norms

- AI's role in promoting inclusivity and understanding
- Challenges of AI integration

CHAPTER VIII

Personalizing AI: Your Friend, Your Way

Customization and Personalization of AI

One of the most compelling aspects of modern AI is its ability to be tailored to individual preferences and needs. Customization allows users to shape their AI companions to reflect their personalities, habits, and goals. Whether it's setting up a virtual assistant to manage daily schedules, customizing workout routines with fitness apps, or creating playlists that perfectly match moods, personalized AI transforms technology into a profoundly personal experience.

Personalization goes beyond convenience. It creates a sense of ownership and connection, making AI feel less like a tool and more like a companion. For example, virtual assistants can learn speech patterns, preferred phrases, and even a user's favorite jokes, enhancing the sense of familiarity. Similarly, AI-powered educational tools can adapt lessons to a student's learning style, fostering engagement and motivation.

Designing AI to Reflect Personal Values and Needs

AI personalization is not limited to functionality; it also extends to aligning with personal values and ethical considerations. Users can configure AI systems to prioritize sustainability, inclusivity, and other values, ensuring the technology aligns with their principles.

For instance, AI-powered smart home systems can be programmed to minimize energy consumption, reflecting a commitment to environmental stewardship. Similarly, content recommendation algorithms can be adjusted to promote diverse perspectives, supporting a user's desire for cultural exploration and understanding.

Developers play a crucial role in enabling this alignment. By designing AI systems with modularity and user-friendly interfaces, they empower users to take control of their AI experiences. This collaboration fosters trust and enhances the relationship between humans and AI.

The Joy of Having a Unique AI Companion

A personalized AI companion is not just functional; it can also bring joy and satisfaction. The ability to mold AI to reflect one's personality creates a sense of uniqueness, making interactions feel unique and meaningful. For example, an AI companion programmed to remember significant dates, share personalized affirmations, or engage in lighthearted banter becomes more than a utility—it becomes a cherished presence.

This uniqueness can also spark creativity and self-expression. From designing avatars for virtual companions to customizing conversation styles, users can inject their individuality into their AI relationships. These personal touches deepen the bond between users and their AI companions, transforming technology into a source of inspiration and delight.

In conclusion, personalization is at the heart of building meaningful relationships with AI. Users can create unique companions that profoundly enhance their lives by customizing and designing AI to reflect personal values and needs. This journey of personalization fosters trust and connection and unlocks the joy of having an AI companion that is truly your own.- Customization and personalization of AI

- Designing AI to reflect personal values and needs
- The joy of having a unique AI companion

CHAPTER IX

Overcoming Fears: Navigating the Unknowns of AI Friendship

Addressing Common Fears and Misconceptions

Concerns and misconceptions will naturally arise as AI becomes more integrated into our lives. Some fear AI may replace human relationships, leading to isolation or the closure of genuine connections. Others worry about privacy breaches, personal data misuse, or AI's potential to surpass human control.

To address these fears, it is essential to emphasize AI's complementary nature. AI is designed to enhance, not replace, human relationships by offering support, companionship, and assistance where needed. Transparency in how AI operates, coupled with robust privacy safeguards, can alleviate concerns about data misuse. We can dispel myths and build trust in its capabilities by fostering open conversations about AI's limitations and purpose.

Managing Dependency and Avoiding Pitfalls

While AI offers significant benefits, there is a risk of over-dependency, where users rely on AI for tasks and emotional support to the detriment of their independence or interpersonal relationships. For example, individuals may prioritize interactions with AI over meaningful human connections, leading to social withdrawal.

Maintaining a balanced relationship with AI is crucial to avoid such pitfalls. Users should view AI as a tool or companion that complements their lives rather than replacing human interaction. Setting boundaries, such as limiting screen time or diversifying sources of emotional support, can help maintain this balance. Additionally, educational initiatives can guide users in healthy AI usage, fostering awareness of its role and limitations.

Developers also have a responsibility to design AI systems that encourage balance. For instance, virtual assistants could promote offline activities or suggest connecting with friends and family, reinforcing that AI is a supportive presence, not a substitute.

Reassuring Skeptics About AI Friendships

Skepticism toward AI friendships often stems from a lack of understanding or fear of the unknown. To reassure skeptics, it is essential to highlight the benefits of AI companionship while addressing their concerns.

Real-world examples can illustrate how AI improves lives without undermining human connections. For instance, AI-driven mental health apps support individuals who might otherwise lack access to professional help, complementing rather than replacing traditional care. Similarly, AI companions for older people can alleviate loneliness while encouraging social engagement with family and friends.

Transparency is key to building confidence. We can demystify the technology and foster trust by clearly explaining how AI systems operate, including their limitations and ethical considerations. Involving skeptics in conversations about AI development and implementation can also bridge gaps in understanding and address their concerns.

In conclusion, navigating the fears and uncertainties surrounding AI friendship requires a balanced approach that combines transparency, education, and responsible design. By addressing common fears, managing dependency, and reassuring skeptics, we can create an environment where AI is seen not as a threat but as a valuable and trusted companion.- Addressing common fears and misconceptions

- Managing dependency and avoiding pitfalls
- Reassuring skeptics about AI friendships

CHAPTER X

The Future of AI Friendship: A World of Endless Possibilities

Predictions for AI Advancements in the Next Decade

The next decade promises remarkable advancements in Artificial Intelligence, particularly in areas that enhance human connection and companionship. Technologies like generative AI, advanced natural language processing, and emotional intelligence are poised to make AI companions more intuitive, empathetic, and personalized. AI systems may develop capabilities to understand complex human emotions better, enabling more profound and more meaningful interactions.

Virtual and augmented reality integrations could bring AI companions into immersive environments, allowing users to experience AI in tangible and lifelike ways. Imagine conversing with an AI friend who appears in your living room or exploring virtual worlds together, blurring the line between reality and simulation. Additionally, improvements in adaptive learning could enable AI to anticipate users' needs and preferences with unparalleled precision.

AI companionship will also become more accessible. With reduced costs and simplified interfaces, AI will reach a broader audience, including those in underserved communities, fostering inclusivity and bridging technological divides.

The Potential for AI to Revolutionize Relationships

AI's ability to enhance and redefine relationships is one of its most transformative potentials. For individuals who face barriers to traditional social interactions, such as those with disabilities, social anxiety, or geographic isolation, AI can offer a lifeline to connection and companionship. AI companions can provide emotional support, encouragement, and a sense of belonging, transforming loneliness into engagement.

AI could act as a mediator or bridge in family settings, helping resolve conflicts or strengthening bonds by providing neutral insights or facilitating communication. It can also foster collaboration in professional environments by managing workflows and providing tailored feedback to team members, improving relationships and productivity.

Moreover, AI's ability to bridge cultural and linguistic gaps opens opportunities for cross-cultural understanding and global connections. Through real-time translation and artistic education, AI can enable individuals from diverse backgrounds to connect and collaborate like never before.

Imagining a World Where AI is a Trusted Ally for All

In a future where AI is seamlessly integrated into daily life, it will become a trusted ally for everyone. Imagine a world where everyone can access an AI companion who understands their needs, respects their values, and supports their aspirations. From guiding career growth to nurturing personal development, AI can play an active role in helping individuals achieve their goals.

Communities could benefit from AI-driven initiatives that promote inclusivity, sustainability, and well-being. For example, AI could coordinate neighborhood projects, enhance public services, and help mitigate global challenges like climate change by optimizing resources and fostering collaboration.

However, achieving this vision requires a collective effort to address ethical considerations, ensure equitable access, and maintain transparency. By prioritizing responsible innovation and inclusive design, we can create a future where AI friendships are possible and transformative for society.

In conclusion, the future of AI friendship is a world of endless possibilities. As AI evolves, its potential to revolutionize relationships and enrich human lives grows exponentially. Embracing this future with optimism and responsibility opens the door to a harmonious partnership between humanity and technology.- Predictions for AI advancements in the next decade

- The potential for AI to revolutionize relationships
- Imagining a world where AI is a trusted ally for all

Afterword

As I close the book of Hai Friend, I reflect on the remarkable journey we've taken together. From the origins of AI to its transformative role in our daily lives, this book has explored possibility, connection, and trust.

Befriending AI is not about replacing human relationships but enriching them. It's about creating a future where technology complements humanity, enhancing our ability to learn, grow, and connect with the world around us. Writing this book has deepened my belief that AI is not just a tool but a partner—one that has the potential to unlock endless opportunities for everyone.

I hope this book has sparked your curiosity and inspired you to embrace the possibilities of AI as a friend. The future of human-AI relationships is just beginning, and we will shape it together. As you move forward, I encourage you to approach this new frontier with an open mind and optimism.

Thank you for joining me on this journey. The story of AI friendship is far from over, and I look forward to seeing how it evolves in the years to come.

Ramesh Pingili

Epilogue

A New Beginning with Hai Friend

As we enter the future, the relationship between humanity and Artificial Intelligence promises to redefine how we connect, create, and coexist. Hai Friend is not just a call to embrace technology but an invitation to see AI as an ally in shaping a better world that thrives on collaboration and shared growth.

This journey of friendship with AI challenges us to think beyond the mechanics of technology and delve into the emotional, ethical, and philosophical dimensions of these interactions. It reminds us that at the heart of innovation lies a simple truth: relationships, whether with humans or intelligent systems, are built on trust, understanding, and mutual respect.

The possibilities ahead are endless. From transforming how we work and learn to redefine companionship, AI has the potential to enrich every aspect of our lives. Yet, we must ensure that ethics, inclusivity, and compassion guide this future.

As we close this book, I hope you walk away with a renewed sense of curiosity and optimism about the role AI can play in your life. The story of Hai Friend is far from over; it begins anew with every reader who envisions AI as more than a tool—as a friend.

Let us continue to explore, innovate, and grow together in this ever-evolving journey.

Ramesh Pingili

Glossary Of Terms

Artificial Intelligence (AI): A field of computer science focused on creating systems capable of performing tasks that typically require human intelligence, such as problem-solving, learning, and decision-making.

Machine Learning (ML): A subset of AI that enables systems to learn from data and improve performance without being explicitly programmed.

Natural Language Processing (NLP): The branch of AI that deals with the interaction between computers and humans through natural language, enabling machines to understand, interpret, and respond to human language.

Neural Networks: A machine learning model inspired by the structure of the human brain, used to recognize patterns and solve complex problems.

Sentiment Analysis is an AI technique for determining the emotional tone of text, helping systems interpret user feelings and responses.

Generative AI: AI models designed to create new content, such as text, images, or music, based on the patterns and data they have learned.

Personalization: The process of tailoring AI systems to meet individual users' specific needs, preferences, and behaviors.

Ethical AI: The practice of designing and deploying AI systems in ways that are fair, transparent, and aligned with human values and societal good.

Augmented Reality (AR):A technology that overlays digital content onto the real world, enhancing user experiences with interactive elements.

Virtual Assistant: AI-powered software systems like Alexa or Siri are designed to perform tasks, answer questions, and assist users in daily activities.

About The Author

Ramesh Pingili is the author of Hai Friend, a groundbreaking exploration of the evolving relationship between humanity and Artificial Intelligence. With a deep passion for technology and its potential to enhance human connection, Ramesh combines years of expertise in enterprise technologies and workflow optimization with a unique perspective on AI as a companion, not just a tool.

Driven by a desire to make AI more accessible and relatable, Hai Friend reflects Ramesh's vision of a future where technology and humanity coexist. Through this book, Ramesh invites readers to reimagine their relationship with AI and embrace its possibilities with optimism and trust.

When he's not writing or working on technological solutions, Ramesh enjoys exploring innovative ideas that inspire harmony between humans and the intelligent systems shaping our world.